AF483839

OCEAN ANIMALS

Ways to use this book

Younger Readers

Read the book together and let your reader hear the clues and see the close-up picture as they guess the animal.

Older Readers

Read each clue aloud without showing the close-up picture. Let your reader try to guess the animal as you read the clues. Show the close-up picture if they need an extra clue.

Make It a Game!

Read the clues aloud and try to guess the animal in as few clues as possible. Make your own clues for each animal after reading the extra facts.

LEARNINGSPARKEDUCATIONALPUBLISHING

THE OCEAN BIOME

BIOME FACTS

All of the animals in this book live in the ocean biome. A biome is a large plant and animal community that has a specific climate. The ocean, or marine biome, is the largest biome on Earth. It covers about 70% of Earth's surface.

The ocean biome includes the Pacific, Atlantic, Indian, Southern, and Arctic oceans. Oceans can be warm or cold. Some animals live close to the surface, and some live deep underwater. All animals in this biome share the trait of thriving in saltwater.

The ocean biome is divided into 5 zones. These zones are organized by the amount of sunlight they receive and water depth. The five zones in order of shallow to deep are: sunlit zone, twilight zone, midnight zone, abyssal zone, and hadal zone.

The sunlit zone extends about 650 feet deep. Sunlight doesn't reach the abyssal or hadal zones. Most of the animals in this book spend their life in the sunlit zone.

I have a **BEAK**.

I have a **SOFT** body.

I have **THREE** hearts.

I have eight **ARMS**.

WHAT
AM I?

I am an OCTOPUS.

OCTOPUS FACTS

The octopus is best known for having eight arms, but you might not know that each arm has a mini-brain. A large nerve cluster in each arm acts like a mini-brain and allows the arms to perform different tasks at the same time. The suckers along each arm are able to smell and taste as the octopus crawls along the ocean floor.

An octopus can swim, but it often prefers to crawl. There's a reason the octopus might prefer crawling rather than swimming, and it has to do with its hearts.

The octopus has three hearts. Two hearts pump blood through its gills. The third heart pumps blood to its organs. When an octopus swims, the third heart stops beating. Swimming makes an octopus tired!

An octopus mostly eats shrimp, crabs, and lobster. It uses its arms to pull the prey toward its beak. The octopus has a venomous beak which paralyzes its prey. If an octopus is threatened, it squirts out a cloud of ink and then zooms away.

I lay **EGGS**.

I can **DIVE DEEP** underwater.

I need to **BREATHE** air.

I have a **SHELL**.

WHAT
AM I?

I am a SEA TURTLE.

SEA TURTLE FACTS

There are several kinds of sea turtles, but they are all reptiles. Since they are reptiles, sea turtles need to breathe air, but they can stay underwater for hours. Sea turtles even sleep underwater!

Most sea turtles stay in the water all the time, and only go on land to lay their eggs. They dig a hole in the sand with their rear flippers. They lay their eggs, called a clutch, then return to the sea.

Baby turtles are about two inches long. After ten years, a sea turtle will be about the size of a dinner plate. The leatherback is the largest sea turtle and can grow up to 8 feet long and weigh 2,000 pounds.

The leatherback sea turtle also dives deeper than other sea turtles. It dives to almost 4,000 feet! The leatherback turtle's soft shell helps it dive so deep. Other sea turtles have hard shells. All sea turtles have one shell feature in common: sea turtles cannot retract into their shells like freshwater turtles.

I have **NO** brain, eyes, or heart.

I am mostly made of **WATER**.

I have a **BELL-SHAPED** body.

I have **STINGING** tentacles.

WHAT
AM I?

I am a JELLYFISH.

JELLYFISH FACTS

Jellyfish are named for their gelatin-like bodies. They are made of more than 95% water and have no brain, bones, teeth, eyes, or heart. The jellyfish has a bell-shaped body and long tentacles. Their tentacles are covered in small, stinging cells. When something touches the stinging cell, a tiny, sharp barb pops out. This barb injects venom into whatever it touches. This paralyzes prey and then the jellyfish can eat it.

The jellyfish's mouth is at the center of its body, and inside its bell are open chambers that act like stomachs. The mouth takes food in, but it also expels waste once food is digested. The jellyfish's mouth can also shoot the jellyfish forward with a stream of water.

Jellyfish mostly eat plankton, but they also eat crabs, fish, and sometimes other jellyfish. Jellyfish do have to worry about being eaten by some fish, and sea turtles love to eat jellyfish.

I have many **TEETH**.

I have **GILLS**.

My skin feels like **SANDPAPER**.

I have a **FIN** on my back.

WHAT
AM I?

I am a SHARK.

SHARK FACTS

There are over 500 species of sharks in the world. The largest is the whale shark, measuring up 60 feet long. The smallest is the dwarf lantern shark, at just 6 inches long.

Sharks have multiple gill slits on each side of their body. Swimming pushes water into the shark's mouth and through its gills. This is like breathing for the shark. Many sharks can pump water through their gills while resting, but some cannot. Sharks that cannot do this have to keep swimming. If water stops flowing through the shark's gills, it will suffocate.

Sharks also have many rows of teeth which they are constantly losing and replacing. Shark teeth vary depending on a shark's diet. Some sharks have sharp teeth, and some have flat teeth.

Sharks have skin made of tiny v-shaped scales. These scales feel smooth from head to tail but feel like sandpaper in the opposite direction.

I can be **BIG** or **SMALL**.

I sometimes **HIDE** in sand.

My tail has a **VENOMOUS** spike.

I am a **FLATTENED** fish.

WHAT
AM I?

I am a STINGRAY.

STINGRAY FACTS

Stingrays are found in fresh water and salt water, but most stingrays live in the oceans. Rays are a type of fish closely related to sharks. They are sometimes called "flattened fish" or "flattened sharks".

Stingrays get their name from the stinger on their tail. The stinger is called a caudal barb and is used for protection. It can stick into a predator and inject venom.

Some rays live in deep water, but most live along the coasts. Stingrays usually swim along the bottom of the ocean, where they hide in the sand. They come in different sizes and can be hard to see in the water. The short-nose electric ray is only 4 inches across and weighs less than a pound. The largest saltwater stingray is the smalleye ray which grows as large as a king-size bed.

People wading in shallow water might step on a stingray and get stung. Doing the "stingray shuffle" when wading in the ocean stirs up the sand and scares away stingrays.

I can only live in **SALTWATER**.

I am **NOT** a fish.

I have no **BRAIN** or **BLOOD**.

I can have **FIVE** arms or up to **FORTY** arms.

WHAT
AM I?

I am a SEA STAR.

SEA STAR FACTS

Commonly called a starfish, the sea star is not a fish. Sea stars are part of an animal group that includes urchins and sponges. There are over 1,600 species of sea stars found all over the world, but they all have one thing in common: they can only survive in salt water.

Some sea stars live close to the shore, in shallow water, or in coral reefs. Other sea stars live in the sand almost 30,000 feet below the surface!

Most sea stars have five arms, but a few species can have up to forty arms. If a sea star loses an arm, it will grow back!

At the end of each arm is a compound eye, called an eyespot. The eyespot sees light and dark, and it can detect movement. It does not see detailed images.

Sea stars do not have a brain, heart, or blood. Instead they have a tangle of nerves called a nerve net, and they pump filtered sea water through their body to absorb nutrients.

I eat **MEAT**.

I live in shallow and deep **WATER**.

I am usually **SHY** and like to **HIDE**.

I have a **SNAKE-LIKE** body.

WHAT
AM I?

I am an EEL.

EEL FACTS

There are about 800 eel species. Some eels live in fresh water, but most are found in salt water. Even freshwater eels usually return to the ocean to lay eggs.

Eels start life as small larvae and mature into snake-like fish. They have one long fin that runs along the top and bottom of their body. This fin allows them to swim forward and backward by waving their bodies one way or the other.

Adult eels are carnivores, meaning they eat meat. The moray eel is unique among eels. It has two jaws. First, it catches prey in its mouth. Then, a second set of teeth extends from inside its mouth. This set of teeth pulls the prey down its throat. Eels eat fish, shrimp, crabs, and just about anything that is smaller than itself. Some eels even eat other eels!

Despite being predators, eels are usually shy and like to hide. They often live in coral reefs or in rock crevices, and they only come out at night to hunt.

Clues by YOU!

It's your turn to create clues. Look at the pictures and read the facts for the following ocean animals. Can you come up with four clues for each animal based on their appearance, behavior, or some other fact?

Tell your clues to a friend or family member and see if they can guess the animal without looking at the picture.

I am a **SEAHORSE.**

Seahorses are tiny fish. They are named for their horse-like shaped head.

A seahorse can change the amount of air in its swim bladder to swim up or down. If a seahorse wants to stay still, it can wrap its tail around something like seaweed.

Seahorses are slow swimmers, but luckily they don't have many natural predators. They are covered in small, spiny plates which makes them difficult to eat.

Male and female seahorses mate for life. The female seahorse puts eggs in a pouch on the male's belly and he later gives birth.

I am a **SPINY LOBSTER.**

The spiny lobster gets its name from the spines that cover its shell. These spines protect it from predators.

The spiny lobster does not have big claws like the common lobster. Instead it has two small claws.

Spiny lobsters live underwater in coral reefs and other places with exposed rock. They hunt at night and eat almost anything.

I am a **SEA ANEMONE.**

There are more than 1,000 sea anemone species. They live both in shallow water and at depths of more than 32,000 feet.

Anemones look like plants, but they are animals. They can move, walk, and swim. Adult anemones often attach themselves to a surface and stay there.

Like jellyfish, sea anemones have stinging cells on their tentacles. Some fish, like clownfish, can live within an anemone's tentacles.